21st Century Junior Library

DISCOVER THE ICHTHYOSAURUS

Josh Gregory

Our Prehistoric World: Dinosaurs

Published in the United States of America by:

Cherry Lake Press
2395 South Huron Parkway, Suite 200, Ann Arbor, Michigan 48104
www.cherrylakepress.com

Content Adviser: Gregory M. Erickson, PhD, Dinosaur Paleontologist, Department of Biological Science, Florida State University, Tallahassee, Florida

Reading Adviser: Beth Walker Gambro, MS, Ed., Reading Consultant, Yorkville, IL

Photo and Illustration Credits: © Michael Rosskothen/Shutterstock, cover, title page; © Dotted Yeti/Shutterstock, 5; Ryan Somma, CC BY-SA 2.0, via Wikimedia Commons, 6; © Yama Hlib/Shutterstock, 7; © Dotted Yeti/Shutterstock, 8-9 © Art Directors & TRIP/Alamy, 10; © Sabena Jane Blackbird/Alamy, 13; © Russell Shively/Alamy, 14; © Dotted Yeti/Shutterstock, 16; © Michael Rosskothen/Shutterstock, 19; © Jack Sullivan/Alamy, 20-21

Cherry Lake Press is an imprint of Cherry Lake Publishing Group.

Library of Congress Cataloging-in-Publication Data has been filed and is available at catalog.loc.gov.

Cherry Lake Press would like to acknowledge the work of the Partnership for 21st Century Learning, a Network of Battelle for Kids. Please visit Battelle for Kids online for more information.

Printed in the United States of America

Cherry Lake Press

CONTENTS

WHAT WAS AN ICHTHYOSAURUS?

When you think about **prehistoric** animals, you probably think of dinosaurs. But dinosaurs were not the only animals living millions of years ago. The *Ichthyosaurus* was another type of prehistoric animal. These ocean **reptiles** lived between 250 million and 65 million years ago.

The *Ichthyosaurus* was among many prehistoric underwater animals.

The name *Ichthyosaurus* means "fish lizard." But *Ichthyosauruses* were not fish. They were not dinosaurs either. They are related to lizards and snakes that live today.

Like dinosaurs, *Ichthyosauruses* are **extinct**. None of them are living today.

The *Ichthyosaurus* may have sometimes jumped out of the water like modern-day dolphins do.

Look!

Look at a picture of an *Ichthyosaurus*. Now look at some fish and lizards. How is an *Ichthyosaurus* like a fish or lizard? How are they different?

WHAT DID AN *ICHTHYOSAURUS* LOOK LIKE?

Ichthyosauruses looked a lot like porpoises or dolphins. They had long bodies with tails. They also had fins and long, pointy noses. Each *Ichthyosaurus* had four fins. Two fins were located on each side of its body.

Can you point to this *Ichthyosaurus*'s tail and dorsal fin? How many fins do you count on one side?

This artwork in Berlin-Ichthyosaur State Park in Nevada shows how big some *Ichthyosauruses* could be.

Some *Ichthyosauruses* were about the size of modern dolphins. They were around 7 to 10 feet (2.1 to 3 meters) long. Others were much longer. The longest one discovered so far was more than 70 feet (21.3 m) long. That is about three-fourths the length of a professional basketball court!

Some *Ichthyosauruses* had very large eyes. Scientists learned this by looking at *Ichthyosaurus* **skulls**. One type of *Ichthyosaurus* had eyes the size of soccer balls! Scientists believe that this *Ichthyosaurus* probably lived in deep water. It is very dark in the deep parts of the ocean. Big eyes helped the *Ichthyosaurus* see.

Think!

Why is it so dark deep in the ocean? Think about where light comes from. What might stop it from reaching deep below the water's surface?

Scientists estimate the size of an *Ichthyosaurus*'s eyes by looking at its eye sockets. Large eye sockets mean the animal had large eyes.

An *Ichthyosaurus*'s long mouth was filled with teeth.

The *Ichthyosaurus* had a long, narrow mouth full of teeth. One type of *Ichthyosaurus* had a short lower jaw. Its upper jaw was long. The teeth on its upper jaw stuck out sideways!

The *Ichthyosaurus* was a master swimmer.

HOW DID AN *ICHTHYOSAURUS* LIVE?

The *Ichthyosaurus* lived in water. As a result, it spent a lot of time swimming. Its tail flipped back and forth. This pushed the *Ichthyosaurus* forward. The fins on its sides helped it change direction. Fins also helped the *Ichthyosaurus* slow down.

Ichthyosauruses were **predators**. They chased after other sea creatures for food. *Ichthyosauruses* used their long mouths to bite down. *Ichthyosauruses* ate mostly ancient sea animals similar to today's squids. Sometimes they also ate fish and other ocean animals.

The *Ichthyosaurus* hunted other animals for food.

Experts are careful not to damage fossils when they study them or put them on display.

Ichthyosauruses needed to breathe air even though they lived in water. This means they had to come up to the surface to breathe. They probably didn't ever leave the water completely, though.

If you want to see an *Ichthyosaurus*, visit a museum. They might have some **fossils** to look at!

Make a Guess!

Scientists study the *Ichthyosaurus* by looking at fossils. What kinds of things can fossils tell us about an animal? What kinds of things are harder to learn from fossils?

GLOSSARY

dorsal fin (DOR-suhl FIN) a fin located on an animal's back

extinct (ik-STINGKT) describing a type of plant or animal that has completely died out

fossils (FAH-suhlz) the preserved remains of living things from thousands or millions of years ago

predators (PRED-uh-turz) animals that live by hunting other animals for food

prehistoric (pree-his-TOR-ik) belonging to a time before history was recorded in written form

reptiles (REP-tilez) cold-blooded animals that have backbones and usually reproduce by laying eggs

skulls (SKUHLZ) the bones that make up an animal's head

FIND OUT MORE

Books

Dahl, Michael. *Monster Fish: The Adventure of the Ichthyosaurus.* Minneapolis, MN: Picture Window Books, 2005.

Korb, Rena. *Discovering Ichthyosaurus.* Edina, MN: Magic Wagon, 2008.

Websites

With an adult, learn more online with these suggested searches.

BBC Nature—Ichthyosaurus

Watch a video that shows *Ichthyosaurus* fossils.

Ichthyosaurus Page

Read more about the *Ichthyosaurus*.

INDEX

ABOUT THE AUTHOR

Josh Gregory writes and edits books for kids. He lives in Chicago, Illinois.